19131

MW00964172

SUSTAINABLE WORLD
ENVIRONMENTS

Rob Bowden

KIDHAVEN
PRESS™

THOMSON

GALE

San Diego • Detroit • New York • San Francisco • Cleveland
New Haven, Conn. • Waterville, Maine • London • Munich

© 2004 by KidHaven Press. KidHaven Press is an imprint of The Gale Group, Inc., a division of Thomson Learning, Inc.

KidHaven™ and Thomson Learning™ are trademarks used herein under license.

For more information, contact
KidHaven Press
27500 Drake Rd.
Farmington Hills, MI 48331-3535
Or you can visit our Internet site at http://www.gale.com

Commissioning Editor: Victoria Brooker
Book Designer: Jane Hawkins
Consultant: Dr. Rodney Tolley
Hodder Children's Books
A division of Hodder Headline Limited
338 Euston Road, London NW1 3BH

Book Editor: Margot Richardson
Picture Research: Shelley Noronha, Glass Onion Pictures

Cover: Flowers blooming on a dry lake bed in the United States.
Title page: Deforestation in Borneo, one of the world's most pressing environmental problems.
Contents page: Zebra and wildebeest in Kenya's Masai Mara.

Picture credits: cover David Woods/Corbis; title page W. Lawler/ Ecoscene; contents David Hosking/ FLPA; 4 Fritz Pölking/ FPLA; 5 Reuters/ Popperfoto; 6 Klaus Andrews/ Still Pictures; 6 (inset) William Gray/ Oxford Scientific Films; 7 M. & C. Denis-Hoot/ Still Pictures; 8 F Lantina/ Minden Pictures; 8 (right) Dominique Halleux/ Still Pictures; 9 F. Bavendam/ FLPA; 10 Dipak Kumar/ Popperfoto; 11 (top) Joe Klamar/ Popperfoto; 11 (bottom) Dean Conger/ Corbis; 12 John Jones/ Papilo; 13 Henryk T. Kaiser/ Rex Features; 14 Mark Edwards/ Still Pictures; 15 Popperfoto; 16 Todd A. Gipstein/ Corbis; 17 (right) Jeremy Sutton Hibbert/ Rex Features; 17 (left) Rex Features; 18 Volodymr Repik/ Popperfoto; 19 (top) Sipa Press/ Rex Features; 19 (bottom) Neil Cooper/ Panos Pictures; 20 Mark Edwards/ Still Pictures; 21 Jeremy Horner/ Panos Pictures; 22 Fred Bruemmer/ Still Pictures; 23 Mark Edwards/ Still Pictures; 24 David Keith Jones/ Images of Africa Photobank; 25 Phil Schermeister/ Corbis; 26 Sue Cunningham/ SCP; 27 (left) David Drain/ Still Pictures; 27 (right) Sally Morgan/ Ecoscene; 28 David Hosking/ FPLA; 29 (top) Rob Bowden/ EASI-Images; 29 (bottom) Nigel Dickinson/ Still Pictures; 30 Harmut Schwarzbach/ Still Pictures; 31 Kathie Atkinson/ Oxford Scientific Films; 32 J. P. Delobelle/ Still Pictures; 33 Julio Etchart/ Still Pictures; 34 Andre Maslennikov/ Still Pictures; Eriko Sugita/ Popperfoto; 36 (top) Topham Picturepoint; 36 (right) David Hoffman/ Still Pictures; 37 David T Grewcock/ FPLA; 38 Gilles Nicolet/ Still Pictures; 39 Rob Bowden/ EASI-Images; 400 Rob Bowden/EASI-Images; 41 Brian Cushing/ Papilo; 42 Mark Edwards/ Still Pictures; 43 Wayne Lawler/ Ecoscene; 44 Roy Maconachie/ EASI-Images; 45 Rhodri Jones/ Panos Pictures.

LIBRARY OF CONGRESS CATALOGING-IN-PUBLICATION DATA

Bowden, Rob
  Environments / by Rob Bowden.
     p. cm. — (Sustainable world)

Includes bibliographical references and index.
  ISBN 0-7377-1898-6 (lib. bdg. : alk. paper)

  1. Sustainable development. 2. Environmental management. 3. Ecosystem management.
  I. Title. II. Sustainable world (Kidhaven Press)
  HC79.E5B69 2004
  337.—dc21

                                                            2003052950

Printed in Hong Kong

Contents

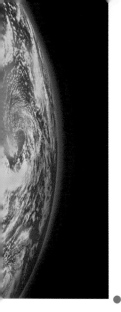

Why sustainable environments?

ALL LIFE ON EARTH DEPENDS ON the environment in which it lives. If the ability of that environment to support life is damaged, or destroyed, then the species living there must move, adapt to the changes, or face almost certain death. Unfortunately, there are many examples on Earth where the process and impact of environmental loss is all too clear to see. The removal of forests, for example, threatens some of the Earth's best-known animals including giant pandas in China, mountain gorillas in Uganda, and orangutans in Sumatra and Borneo.

A lioness captures a zebra on the African plains as part of a continual cycle of life that has been sustainable for thousands of years.

CYCLE OF LIFE

An environment and its species are known as what scientists call an ecosystem. Within any ecosystem species depend upon one another to a greater or lesser extent. In the savannah grasslands of Africa, for example, lions depend on species such as zebra, wildebeest and impala for their prey. In turn the feces of the lions help to fertilize the grass that its prey feeds on. In this way ecosystems can be thought of as circular or as cycles. An ecosystem can be any size from a garden pond to the entire Earth.

Massive population growth has placed natural environments and wildlife under greater pressure than ever before.

OPINION

'The twentieth century created such big pressure on the biosphere, on the environment, that we can now speak of the real environmental crisis in the world. The problems of the environment will compel us to change our way of life, our values, our goals, our technologies.'

Mikhail Gorbachev, former leader of the Soviet Union and head of Green Cross International

DISTURBING BEHAVIOR

Humankind as we know it today has only been on the Earth for around five hundred thousand years, but in that time humans have manipulated ecosystems and environments more than any other species. At the start of the twenty-first century, as human populations continue to grow, they are disturbing the Earth's natural cycles more than ever. For example, humans often remove parts of the ecosystem for their own use and transport them over great distances so that they will never return to their local origins. The removal of fish from the oceans, or of timber from tropical forests, are examples of this.

Human impact on environments and wildlife, if allowed to continue unchecked, could result in disastrous consequences for all life on Earth, as ecosystems collapse along with the life they support. Sustainable development offers a solution to this problem. It promotes the use of environments and wildlife for human benefit, but in ways that ensure they are preserved for the benefit of future generations too. This book will explore how sustainable development can help us to work with environments and wildlife and not against them.

Environmental problems

· ·

LOOKING AT THE ENVIRONMENTAL PROBLEMS of the Earth is a complicated task. This is because the problems are many and varied. They are also interconnected, meaning that they tend to affect one another. For example, the burning of fossil fuels (coal, oil, and natural gas) to provide energy, releases vast quantities of carbon dioxide (CO_2) into the atmosphere. This additional carbon dioxide traps the sun's heat, causing a warming of the climate that is in turn melting the polar ice caps. This warming is also causing sea levels to rise, a factor that is made worse by the melting polar ice. Many of the Earth's coastal ecosystems are threatened by the rising sea levels.

Below: Ice and snow in polar regions show signs of melting as the Earth's climate warms.
Inset: Coastal ecosystems, such as this mangrove forest in the Bahamas, are threatened by rising sea levels.

INTERDEPENDENCE

The Masai have long herded cattle in the Masai Mara, sharing the environment with its native wildlife.

Similar patterns of cause and effect are found at a more local level as well as globally. In China, extreme flooding along the Yangtze River is closely related to deforestation in the upper reaches of the river. Without trees to absorb the rain and protect the soils, the rainwater passes rapidly downstream in destructive surges that threaten natural habitats, and the people and wildlife living there. At whatever scale environmental problems are considered, it is clear that life on Earth is extremely interdependent. In searching for sustainable solutions to environmental problems it is important to remember this.

HUMAN NEEDS

It is also of great importance to remember that people depend on environments too. In parts of Africa, for example, people have, in the past, been excluded from using their surrounding environments because they are also home to some of Africa's more endangered wildlife. Although this might help to protect wildlife, it can cause hardships and suffering for the people living nearby and create anger and resentment. In Kenya, for example, the Masai killed large numbers of wildlife living in Amboseli National Park during the 1970s to protest being banned from using land that had been theirs for many generations. Park authorities had to back down and allow the Masai to continue using parts of their land. The Masai now share in the benefits of wildlife conservation such as the earnings generated by tourists who come to see the wildlife. This example shows that for sustainable development to truly benefit environments and wildlife, human needs must also be considered.

Above: The rosy periwinkle is a rainforest plant from Madagascar that has helped humans to treat life-threatening diseases such as leukemia.
Left: Lowland rainforests, such as here in Borneo, are the richest ecosystems on Earth.

ENVIRONMENTAL HOT SPOTS

The impact of human activities on ecosystems means that environmental problems are found virtually everywhere, including in some of the most remote locations on Earth. However, there are certain locations that face greater challenges than most; these have become known as environmental hot spots.

Arguably the best-known of the environmental hot spots are the world's remaining tropical forests. Today, they cover around just 7 percent of the Earth's land surface, having been reduced by half in the last hundred years. As world population continues to grow and places ever-greater pressure on tropical forests, they are being cleared at a rate of around 1 percent per year.

Tropical rainforests contain at least half of all-known plant and animal species as well as many that are yet to be discovered. Many of these species are of great importance to humans as well as in their natural environment. For example, these forests offer some of the best hopes for finding a cure for HIV/AIDS — a currently incurable disease that kills over 3 million people every year. Many food products including cocoa, coffee, cashew nuts and bananas also originate from forest plants.

CORAL REEFS

Other environmental hot spots include coral reefs; 60 percent of which are threatened by human activities and 11 percent of which have already been destroyed. Coral reefs are the most species-rich ecosystem in the oceans. They contain around a quarter of all-known marine fish species, and may be home to over a million species in total, even though they make up just 0.2 percent of the ocean area. Coral reefs are under threat from tourism, coastal development, pollution, fishing and the mining of coral. But the biggest threat comes from bleaching — a process in which warmer oceans (as a result of climate changes) cause the coral to die.

A coral reef in Papua New Guinea provides a habitat for an incredible variety of oceanic species, many of them endangered.

A farmer in Punjab, India, crosses parched paddy fields normally green with rice.

OPINION

There is new and stronger evidence that most of the warming observed over the last 50 years is attributable to human activities.

United Nations Intergovernmental Panel on Climate Change (IPCC)

A GLOBAL CRISIS?

If there is one environmental problem that shows the global nature of the environment better than any other, it is the problem of climate change. This is the result of global warming caused by our present day lifestyles. Global warming is caused by a blanket of gases around the Earth that trap heat which would otherwise escape into space. This blanket of gases helps keep the Earth warm and acts as the Earth's climate regulator. Without it the Earth would be a frozen, lifeless planet about 86°F colder than it is today. The problem is that human activities, such as the burning of fossil fuels and the removal of forests, have dangerously altered the content of the gases in our atmosphere. In particular, there has been a rise in the level of so-called greenhouse gases which include methane, chlorofluorocarbons (CFCs), and especially carbon dioxide (CO_2). These greenhouse gases have significantly increased the warming of the global atmosphere and lead to changes in the global climate.

Changes in climate patterns are affecting environments and the wildlife that depend on them. Since global temperature records began in 1867, scientists have measured the ten warmest years as all having occurred since 1980, with 1998 the warmest so far. There is also evidence to suggest that global warming is accelerating. Experts warn that by 2100 average temperatures could have increased by as much as 42°F.

LIFE ON A WARMER PLANET

What effect this increase in temperature would have is not entirely clear, but there are already signs of what may lie ahead for life on a warmer planet. Climates are certain to change with warmer temperatures. On the whole, the weather is likely to become wetter across much of the world as higher temperatures accelerate evaporation rates. Although some places, such as southern England, are predicted to become drier. Extreme weather events such as storms, flooding, and droughts are likely to become more frequent, severe and widespread than in the past. If sea levels rise, as predicted, by almost three feet by 2100, one of the biggest changes will be the loss of low-lying environments. Coastal ecosystems, and the people living there, will be especially at risk. In Bangladesh and Egypt, for example, large areas of land, including major human settlements, could find themselves completely submerged.

---
weblinks

For more information about the World Meteorological Organization, go to www.wmo.ch

---

Right: Unusually wet weather led to the flooding of Prague, Czech Republic, in 2002 and the collapse of this building.
Below: This fertile farmland in the Nile Delta of Egypt could soon be permanently flooded due to sea level rises caused by global climate change.

BEYOND CAPACITY

One way to look at the state of the environment is to consider its carrying capacity. This is simply the total population that an ecosystem can support sustainably — without becoming degraded or damaged. If the carrying capacity of an ecosystem is known then it is possible to assess its health accordingly. Using this idea, some scientists have suggested that the Earth's carrying capacity is only 3 billion people, meaning that we are already at over twice our sustainable limit. However, others suggest that the carrying capacity could be as high as 44 billion.

The reality is probably somewhere in between, but any consideration of carrying capacity is made harder by the fact that not all people have the same impact on environments and wildlife. Consider emissions of CO_2, for example, the main greenhouse gas responsible for global warming and climate change. The United States is responsible for 24 percent of global CO_2 emissions and yet it accounts for only 4.7 percent of the world population.

Poor rural households such as this one in Sikkim, India, have a relatively low impact on environments and wildlife.

These elaborate houses in San Francisco, California, can have a significant effect on even distant environments.

BIG FEET, LITTLE FEET

The different impact of individual nations or regions on the environment can be measured by their ecological footprint. This measures the area of land and sea (known as a global hectare, approximately 2.5 acres) needed to provide a population with the food, materials, and energy it needs and to absorb the wastes that it produces. Scientists have used this method to calculate that an ecological footprint of 1.9 global hectares per person is sustainable on a global scale, but this is already exceeded by many nations. The United States, for example, has an ecological footprint of 9.70 global hectares per person and in the United Kingdom it is 5.35. By contrast, the ecological footprint of many less developed countries is currently below the sustainable level at 1.09 in Kenya, and just 0.77 in India.

The challenge for sustainable development is to find ways that will allow poorer countries to enjoy an improved quality of life without extending their ecological footprints beyond a sustainable level. At the same time wealthy nations must act to reduce their ecological footprints.

Toward sustainable environments

CREATING A FUTURE IN WHICH ENVIRONMENTS and wildlife are sustainably managed and developed is a challenging task, but it is nothing new. In fact, efforts to encourage the sustainable management of environments and wildlife have been taking place now for over fifty years. In that time some of the groups at the forefront of these efforts, such as Greenpeace, Friends of the Earth, and World Wildlife Fund (WWF), have become household names and begun influential movements for change.

Pesticides are sprayed on peaches in Italy. Pesticide damage to the environment was highlighted in the book *Silent Spring*.

WARNING SIGNS

Of the many groups and individuals who have contributed toward the more sustainable management of our environments, some have given us vital warning signs of how human actions affect environments and wildlife. One of these warnings came in 1962 from American ecologist and writer Rachel Carson. She published a book called *Silent Spring* that warned the world about the damage done to environments and wildlife by the use of chemical pesticides. The title describes the way in which normally noisy spring mornings, full of bird song, were falling silent as the bird population, and the food it depends on, was being poisoned by the use of chemicals in the environment. To

Friends of the Earth supporters in Hong Kong campaign for cleaner air in urban environments.

many environmentalists, Carson's book marks the point at which concern for environments and wildlife became of wider interest to the general public. It sparked what became popularly known as the environmental movement.

THE PIONEERS

Like any new movement, the environmental movement had its early pioneers, many of whom continue to shape its direction and activities today. They include organizations such as the conservation group WWF, formed in 1961 amid media reports about the fast disappearing wildlife of Africa. WWF has since grown into a worldwide organization operating in 90 countries and supported by some 5 million individuals and businesses. Perhaps the earliest pioneer, however, is the Sierra Club, which was founded in 1892 to encourage the protection and enjoyment of natural environments in the United States. It was from within the Sierra Club that two of the world's best-known environmental groups were started: Friends of the Earth and Greenpeace. Both were established in 1969 by former members of the Sierra Club, who left to encourage more urgent and international action against some of the Earth's major environmental threats.

> ### OPINION
> Over increasingly large areas of the United States, spring now comes unheralded by the return of the birds, and the early mornings are strangely silent where once they were filled with the beauty of bird song.
>
> *Rachel Carson,* Silent Spring, *1962*

— **weblinks** —

For more information about WWF International, go to www.wwf.org

Earth Day celebrations are held every year on March 21st. This day has been celebrated officially by the United Nations since 1971.

AN INTERNATIONAL FOCUS

In the 1970s, the awareness of environmental issues that had been created by various national and regional movements began to take an international focus. In 1971, for example, both Friends of the Earth and Greenpeace expanded to form an international network of concerned environmentalists beyond their North American origins. Another key event was the introduction of Earth Day, an annual day to remind people that they must care for the world they inhabit. Celebrated on March 21 every year, Earth Day was launched in 1970 in San Francisco and officially adopted by the United Nations in 1971.

UNEP

The United Nations (UN) became even more involved when it established the United Nations Environment Program (UNEP) in 1972 at the UN Conference on the Human Environment, held in Stockholm, Sweden. From its headquarters in Nairobi, Kenya, UNEP today provides global leadership toward more sustainable environmental management, by encouraging greater cooperation between the environmental organizations, national governments, and local communities involved. It has helped reach several international agreements aimed at reducing human impact on environments and wildlife. For example, it was the main force behind a 1987 agreement to ban substances that were destroying the ozone layer, causing a so-called ozone hole to appear over Antarctica, which also contributes to global warming.

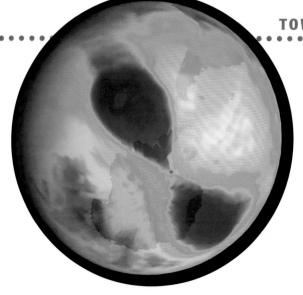

The ozone hole over Antarctica shows up dark blue in this satellite image of Earth.

the media could also be used to raise awareness of their concerns. Greenpeace became masters of this approach with their high profile anti-whaling campaigns, where they used fast inflatable boats to get between whales and the ships trying to hunt them.

Greenpeace protesters use inflatable boats to disrupt Japanese whaling operations, helping to attract media attention for environmental issues.

MEDIA INTEREST

At the same time as environmental organizations were becoming more international in their focus, so too was the world's media. Improved technology allowed events in distant locations to be beamed into the homes of people across the world. This greatly helped to reinforce the message of environmentalists that there was Only One Earth and that we should all be concerned about environmental issues, whether they be on our doorstep or on the other side of the world. Issues such as deforestation in the Amazon rainforest and dwindling populations of African elephants due to poaching began to make world headlines. This publicity strengthened the position of already popular environmental campaign groups. Some groups quickly learned that

GAINING GROUND

During the 1980s, the environmental movement began to gain rapid ground. The environment was now firmly established, not just as an issue for scientists, but as a concern for the general public. Millions of people signed up to support environmental groups, school projects on environmental issues became widespread, and universities began to offer specialist courses in environmental studies or management. Even celebrities were becoming involved by promoting fund-raising campaigns for the protection of endangered species or habitats. Support for environmental issues also came from the increased leisure and travel that many people, particularly in more developed countries, were beginning to enjoy. Many chose to participate in activities or holidays associated with preserving natural environments or wildlife, and in doing so became more aware of the problems facing them.

BREAKING POINT

The 1980s also saw more high-profile environmental issues in the media. In 1986, for example, a nuclear reactor at Chernobyl in the Ukraine, exploded, sending clouds of radioactive gases across much of northern Europe. The human and environmental effects of the Chernobyl accident led to concern about building new nuclear power stations in Europe. Another high profile incident was the Exxon Valdez oil spill of 1989. Nearly 11 million gallons of crude oil were spilled into Alaska's pristine waters, contaminating around

The explosion of the Chernobyl nuclear reactor in the Ukraine led many countries to rethink their policy on using nuclear energy.

2,500 miles of coastal ecosystems and killing enormous numbers of wildlife. This disaster again highlighted the damage human actions were having on many of the world's most fragile environments.

The plight of humans who were affected by the degradation of environments was also demonstrated during the 1980s. In Sudan and Ethiopia, for example, about eight hundred thousand people were seen on television starving to death in one of Africa's worst famines. Although the causes of the famine were numerous, the degradation of farmlands due to deforestation and overgrazing were considered to be a major factor.

Workers struggle to clean the once pristine beaches in Alaska following the *Exxon Valdez* oil spill.

A refugee camp in Ethiopia is home to thousands of people whose land is no longer able to support them. Environmental refugees, as they are known, are on the rise worldwide.

APPROACHING THE LIMITS

Toward the end of the 1980s, many people believed that the environment was being stretched beyond its limits. As the world population passed 5 billion in 1987, there were signs of environmental stress at almost every turn. Forests and lakes were shrinking, animal populations collapsing, human livelihoods suffering, and global systems breaking down. As the Earth appeared to be near breaking point, it was clear that something had to be done, and soon.

The Tree of Life became a global symbol for the environment at the Earth Summit in 1992.

Earth's environments and wildlife. It also recognized that millions of the world's people lived in extreme poverty, and depended on the environment and its resources for their survival. According to the report, the two issues could not be separated, but instead must be addressed together through the idea of sustainable development. The report defined sustainable development as meeting the needs of the present population without compromising the ability of future generations to meet their own needs.

SUSTAINABLE DEVELOPMENT

Though not the first, or only, report to promote the idea of sustainable development, "Our Common Future" succeeded in making the idea popular. Within just a few years it was being widely used, not just by environmentalists, but by politicians, journalists, teachers, industrialists, and businesses too. Interest in sustainable development was further strengthened at the 1992 Earth Summit in Rio de Janeiro, Brazil. The Earth Summit brought together world leaders, community groups and environmentalists in one of the world's biggest gatherings to discuss the relationships between humans and environments

A NEW AGENDA

In 1987 a report called "Our Common Future" was published, which introduced the world to the idea of sustainable development. The report was written by the World Commission on Environment and Development (WCED), a group asked by the United Nations to produce a global agenda for change. It was different to previous reports because it went beyond recognizing the threats posed to the and wildlife. New international agreements were reached on reducing climate change, protecting biodiversity and, most importantly, on an action plan called Agenda 21, which was designed to support and encourage sustainable development.

START, STUMBLE...

Despite the optimism that surrounded the 1992 Earth Summit, progress toward the sustainable management of environments and wildlife has been slow in the years since the summit. Environmental trends (such as deforestation and soil erosion) have worsened and many governments have failed to meet the various targets they agreed to. At the same time, conditions for millions of the world's poorest people have remained unchanged, and in many cases have deteriorated. In 2002, the World Summit on Sustainable Development (WSSD) in Johannesburg, South Africa, provided an opportunity for reflection on progress since 1992, and a chance to learn from the lessons of the last decade. The summit showed that there was still much to be done and that the need for action had become more urgent than ever.

OPINION

WSSD – The World Summit of Shameful Deals...has failed dramatically to take the action needed to reduce the patterns of unsustainable development and consumption that are impoverishing our planet and the people who live on it.

WWF website

This slum in Calcutta, India, shows that living conditions have little improved for millions of the world's poor since the 1992 Earth Summit.

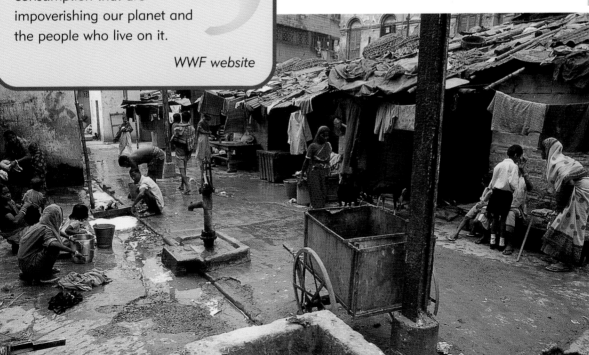

Sustainable environments in practice

MANY EXPERTS BELIEVE THAT the twenty-first century will be a make-or-break century when it comes to the state of the environment. If environmental warning signs are ignored then conditions for people and wildlife are likely to worsen considerably. This need not happen, however. It is possible to create a future in which people, wildlife, and environments all prosper. Examples from around the world demonstrate how this might be achieved and, more importantly, that it is already happening.

This wind farm in Denmark was set up by nearby residents. As well as using the power themselves, they sell it to electricity companies.

THE POWER OF NATURE

Energy consumption is one of the major causes of environmental problems on Earth, ranging from deforestation, to oil spills, to global climate change. Energy is essential for human development as it powers industries, offices, schools and homes around the world. The problem is that much of the world's existing energy comes from fossil fuels such as oil and coal, or is extracted from forests at a rate faster than forests re-grow. It is possible, however, to meet the growing

This solar thermal power plant in California's Mojave Desert uses the sun as a renewable energy source to generate electricity.

global energy demand, and protect environments and wildlife, by switching to sources of energy that are more sustainable. Many of these are available in a virtually limitless supply from natural sources such as the wind, ocean tides, and radiation from the sun. The sources are known as renewable forms of energy. By encouraging greater use of sustainable energy, and becoming more energy efficient, emissions of climate changing gases could be significantly reduced, helping to reduce environmental degradation and wildlife losses across the planet.

RENEWABLE ENERGY

The technology to harness (use for human benefit) renewable sources of energy is now widely available and many countries are beginning to use the sustainable power of nature. In Denmark, for example, about 13 percent of electricity is generated from wind power, while neighboring Germany has grown to be the world's biggest wind energy nation since introducing the technology in the early 1990s. Germany is also a world leader in the development and use of solar energy, along with Japan and the United States. The annual worldwide production of photovoltaic cells (PV cells), that convert the sun's energy into electricity, grew by an incredible 625 percent between 1990–2000.

Despite these positive developments, sustainable energy still only provides a fraction of the world's total energy consumption. Wind energy, for example, accounts for less than 1 percent of global electricity demand.

The Bamburi crocodile farm in Kenya is located within a restored quarry.

restoring its quarries into natural habitats since 1971. They have planted over 2.5 million trees in and around their existing and abandoned quarries. The abandoned quarries are today a series of reserves that teem with wildlife such as hippos, buffalo, giraffes, antelopes, and hundreds of bird and plant species. They have become a popular attraction known together as The Baobab Adventure, and attract thousands of tourists from the nearby resorts of Mombassa. Facilities such as a snake and reptile park, crocodile farm, and forest walking and cycle trails, benefit tourists, but also help to educate them about the importance of environments and their wildlife.

weblinks

For more information about The Baobab Adventure, go to www.kenyabeach.com

RESTORING THE ENVIRONMENT

Although media reports and environmental campaigns often focus on the damage done to environments and wildlife, there are examples of governments, businesses, and communities working to restore damaged environments. In Kenya, for example, the Bamburi Cement company has been

WETLANDS CONSERVATION

Water systems, such as lakes, rivers and wetlands, are particularly vulnerable to human interference. For example, in Denmark, modifications to the flood-plains of the Skjern River led to a 98 percent reduction in the river's wetland area. As a result, wildlife numbers fell dramatically, including the complete disappearance of otters and the near collapse of Denmark's wild salmon population. In 1999 a project was launched by the Danish National Forest and Nature Agency to restore the wetlands and bring back the wildlife that once thrived in the area. Flood embankments were removed so that the river could again meander and flood over the surrounding land. The results of this project are pending, but if successful it will be one of the biggest environmental restoration projects in northern Europe.

A similar project is underway at Lake Tahoe in California, where wetlands were altered in the 1950s to create Tahoe Keys, a development of luxury homes. The alterations led to greater levels of nutrients entering the lake and turning its crystal clear waters cloudy. This was due to eutrophication where the nutrients encourage rapid plant growth and decomposition that starves the water of

oxygen and turns the water murky. Concerned about the effect on Lake Tahoe's wildlife, and the 4 million tourists who visited it each year, local businesses are now working with the community to restore the natural wetlands that filter out the nutrients.

The environment around Lake Tahoe, California, is being restored to its natural conditions following years of damaging urbanization.

> ## OPINION
>
> The public, business owners and government officials…
> realized that for Lake Tahoe to have a healthy economy it must have a healthy ecology.
>
> *Will Hart, Environmental News Service*

We've got trained lecturers here, employed from the local community. They may not know the scientific name of each plant, but they know the common name and the important thing is that they get through to the people, because they speak their language.

Marco Lucena, Administrator,
Guarapiranga Ecological Park

A tourist enjoys the lush vegetation of Guarapiranga Ecological Park, Brazil.

URBAN ENVIRONMENTS

Environmental restoration is not always about rural or wilderness areas. There are important environments in urban areas too and these are just as needy of care and attention. A good example of urban environmental restoration comes from Sao Paulo, Brazil, one of the world's biggest cities. In 1989 the city authorities established the Guarapiranga Ecological Park on the outskirts of the city. The land forming the park had suffered years of deforestation by homeless families who used the timber to build homes in the surrounding shanty towns, or *favelas* as they are known in Brazil. The Guarapiranga reservoir was also suffering from the dumping of raw sewage and other pollutants into its waters.

Tree felling is now banned in the park and since 1995 over a half million trees and shrubs have been planted to restore the park's forest cover. The most fragile areas of the park are strictly protected, but sections are open to the public who use the land for recreation or for controlled fishing. Forest police patrol the park to ensure that there is no illegal hunting, fishing or logging. The park also offers educational tours and receives busloads of Sao Paulo's children who come to learn all about the park and its wildlife.

GREEN SPACES

Urban environments need not be as large as Guarapiranga Ecological Park. Even small areas of wilderness can support a surprising variety of wildlife in and around urban centers. In Britain, for example, the narrow strip of land to either side of its canal network often contains a rich variety of wildlife, even as it passes through some of the biggest industrial cities. These strips act as a sort of wildlife corridor that connects environments that may have been joined up in the past, but are today separated by human activities and settlements. Urban centers, in particular, can present a major obstacle to wildlife and so preserving or restoring green spaces, such as parks and riverbanks, is very important. The benefits also extend to the people living there. Trees, for instance, can help to clean the air and parks and waterways can provide recreational facilities and give urban populations the chance to enjoy wildlife.

Left: A wooded canal provides a wildlife refuge in the heart of industrial England.
Below: Green spaces such as Central Park in New York are important for both wildlife and urban populations.

PROTECTED AREAS

Restoring environments is an expensive and time-consuming process. It is also unlikely to ever return environments to their truly natural state. Many environments have taken thousands of years to develop and cannot be easily recreated by human actions in a matter of just a few decades. A better approach is to protect remaining environments and wildlife before they become degraded. Today most countries have at least part of their land and/or ocean area set aside as protected areas.

Some protected areas are very specific and protect a particular habitat or endangered species. For example, in India, the government has established twenty-three

reserves since 1973 as part of the Project Tiger program to protect the endangered Bengal tiger. There are now fewer than five thousand Bengal tigers left in the wild, down from over one hundred thousand just one hundred years ago. Other protected areas cover vast areas of land and contain a wide variety of wildlife and habitats. Some of the best-known are the parks and reserves of East Africa, such as the Masai Mara, the Serengeti and Tsavo. These enormous areas contain some of the greatest concentrations of wildlife found anywhere on Earth.

India's endangered tiger populations are today better protected due to a number of reserves that conserve the tiger's natural habitat.

DATABANK

By 2000, the world's thirty thousand protected areas covered 8,000,000 km^2, a land surface roughly the size of India and China combined.

Above: Topi are one of many species that make up the world's biggest concentration of mammals in the Masai Mara and Serengeti grasslands of East Africa. Right: A woman picks a medicinal plant used to treat fever from a forest in Sarawak, Borneo.

Protected areas sometimes come under criticism for ignoring the needs of people living around them. As a result, many now allow people to use the land and resources of protected areas providing it is done sustainably. In many forest reserves, for example, people are allowed in to collect fallen branches for use as fuelwood, or to collect forest products that are used for local food or medicines.

Protected areas have also been criticized for creating isolated areas of protection. The concern is that with the creation of protected areas people may ignore environments and wildlife that lie beyond their boundaries. This can lead to protected areas becoming isolated enclaves that are cut off from the surrounding environment and limit the natural movement of wildlife. In India, for example, tigers are unable to move between the reserves because the area in between them is heavily populated and farmed. In Central America a project called The Way of the Panther aims to avoid this problem by linking together reserves in eight countries from Mexico to Panama with planted regions (corridors) in between them. If successful then similar schemes are likely to follow in other regions.

---
weblinks

For more information about the world's protected areas, go to The World Conservation Union's website at www.iucn.org

---

29

VALUING NATURE

Trekkers enjoy the natural splendor of the Annapurna mountains in Nepal, but their interest can also leave its mark.

One of the most effective ways to encourage the sustainable management of environments and wildlife is to help people realize the value of nature. Where people recognize the value of an environment, they are more likely to act to protect it. Tourism is one of the main ways in which the value of environments and wildlife can be seen. As travel has become easier and cheaper, people increasingly look to remote regions of the world for their travel experiences, and visiting environments and wildlife are often high on their list of priorities. Popularity in trips that focus on environments and wildlife offer experiences ranging from whale watching and coral diving to driving safaris and mountain trekking.

Not all tourism is beneficial, however. If poorly managed, visitors can disrupt wildlife and contribute to environmental degradation. In Kenya's Masai Mara, a 250 percent increase in tourist numbers between 1986 and 1998 has placed pressure on some animal populations. Cheetahs, for example, have changed their hunting patterns to avoid the mass of tourists, but this means hunting in the middle of the day, when the heat makes hunts less successful and so threatens their very survival. In other regions, such as the Himalayas, the popularity of mountain trekking has led to problems of litter, footpath erosion, and deforestation.

ECOTOURISM

A more responsible form of tourism, known as ecotourism, has grown in popularity in recent years. Ecotourism is intended to respect local environments, and the people and wildlife living there, and in doing so make a valuable contribution to sustainable environmental management. By involving and benefiting local communities, ecotourism can also help to encourage greater care for environments and their wildlife as people realize the value of the tourism that they attract. In Ecuador, for example, the Achuar people of the Amazon are now heavily involved in ecotourism through lodges such as Kapawi. Previously the Achuar were clearing the forest for cattle ranching, but today they earn around 45 per cent of their income by providing goods and services to ecotourism in the region. Once the Achuar people are fully trained in managing Kapawi Lodge it will be given to the community for their own benefit in 2011. It is hoped that by then it will be a sustainable project for both the people and the environment.

A guided reef walk on Australia's Great Barrier Reef at low tide is typical of the many activities that are known as ecotourism.

weblinks

For more information about the Great Barrier Reef, go to www.gbrmpa.gov.au

OPINION

The trekker's demands for heating and hot water have led to increased deforestation, there are litter and sanitation problems, and wildlife has been driven away from many populated areas.

Nepal Study Tours and Treks website

WORKING WITH WILDLIFE

It is not possible to set aside all environments for the benefit of wildlife, as humans also need to use the environment for their own advantage and survival. There are, however, ways of working with environments and wildlife for mutual gain. In farming, for instance, there are numerous techniques that both conserve environments and wildlife and contribute to higher crop yields and a better standard of living for farmers.

AGROFORESTRY

Agroforestry, a technique of combining trees with crops, is a good example of sustainable development. The trees help to reduce soil erosion by binding the soil together with their roots, and their canopy helps reduce the eroding force of winds. Trees provide shade for livestock and people, and for crops that might otherwise be difficult to grow. Some trees, such as the Neem tree, actually help to fertilize the soil by absorbing nitrogen from the air and passing it into the soils. The Neem also deters insects, including mosquitoes, that in many tropical areas carry malaria, a potentially deadly disease for humans. Other tree species can provide fruits, nuts, or medicinal products for human use, and the leaves of several species can be used as feed for livestock. If carefully managed,

Here in the Dogon region of Mali, farmers grow crops underneath the shade of trees. This is one of many agroforestry techniques that helps to conserve natural environments.

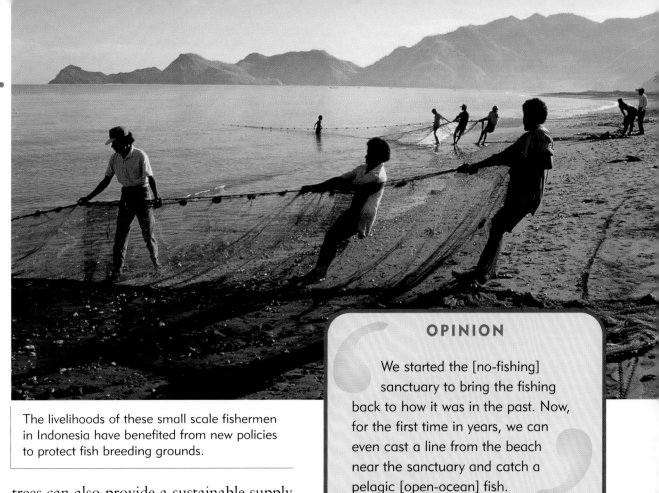

The livelihoods of these small scale fishermen in Indonesia have benefited from new policies to protect fish breeding grounds.

OPINION

We started the [no-fishing] sanctuary to bring the fishing back to how it was in the past. Now, for the first time in years, we can even cast a line from the beach near the sanctuary and catch a pelagic [open-ocean] fish.

Dolvi Janis, Blongko village head,
Indonesia

trees can also provide a sustainable supply of fuelwood. While providing these human benefits, agroforestry preserves habitats for a wide variety of wildlife and particularly benefits bird populations that often decline rapidly when tree cover is removed.

NO-FISHING ZONES

In the oceans, fishermen have found that by working with nature they can significantly improve their fish catch while at the same time conserving biodiversity. In Indonesia, for example, several fishing communities have begun to set aside areas of the nearby coral reefs as no-fishing zones. Within just a few years the communities began to notice the benefits.

These included a greater variety and number of fish beyond the no-fishing zones, and an increase in the attractiveness of the reefs for other income-generating tourist activities such as snorkeling. While the Indonesian examples are small scale, the same principle can be used on a larger scale. This would not only help to make the world's fishing industry more sustainable, but more than likely lead to greater catches too, proving the benefit of working with environments and wildlife.

Creating sustainable environments

● ●

Managing environments and wildlife sustainably requires social and political support from governments, organizations, and individuals alike. This support can take many different forms and its effectiveness can vary considerably. The challenge for the future is to learn from successful examples and apply them to other issues in other regions.

INTERNATIONAL COOPERATION

Many environmental issues are regional or even global in scale and so require international cooperation to solve them. One of the earliest international environmental agreements was over the problem of acid rain. This is caused by emissions of sulphur dioxide (SO_2) and nitrogen oxides (NOx) that are released as fossil fuels when burned. Once in the atmosphere they mix with water vapor and fall to the ground as acid rain, poisoning lakes and killing trees. Because weather patterns carry the rain over great distances, acid rain becomes an international problem with emissions from one country affecting

These trees in Krkonose National Park, Poland, have been severely damaged by acid rain.

In 1997, politicians from 160 countries met in Kyoto, Japan, to agree on a plan to tackle climate change.

environments in neighboring countries. In 1979, thirty-four European and North American coun-tries agreed to reduce emissions, but progress was mixed because no targets were set. Then, in 1983, Germany and the Scandinavian countries, which were among the worst affected by acid rain, persuaded twenty-one European nations to agree to targets to reduce emissions of SO_2 by 30 percent below their 1980 levels by 1993. By 1988, twelve countries had met this target, but in the same year it was replaced by a new target to reduce emissions by 58 percent by 2003.

KYOTO PROTOCOL

Several other international calls to action have followed since those on acid rain. One of the biggest of these was the 1997 Kyoto Protocol. This stated that countries should aim to reduce emissions of carbon dioxide and other greenhouse gases by 5 percent below their 1990 levels by 2012. Putting the Kyoto Protocol into practice, however, has demonstrated the difficulty of international agreements. Several less developed countries object because they

DATABANK

The Intergovernmental Panel on Climate Change says that a 60 percent cut in emissions will be needed to stabilize carbon dioxide levels in the atmosphere, much higher than that agreed in Kyoto.

produce only a fraction of the gases concerned. They also claim that reducing their emissions could slow economic growth and worsen the already extreme poverty suffered by many of them. The United States and Australia, two of the world's most developed economies, have also refused to commit to the Kyoto Protocol so far. They believe that it could harm their economies and lead to job losses. Their resistance is particularly problematic as they make up the vast proportion of carbon dioxide emissions. The United States for example, accounted for 36.1 percent of carbon dioxide emissions in 1990.

Left: The headquarters of the U.S. Environmental Protection Agency, founded to protect environments and wildlife.
Above: Reducing pollution from factories such as this one in South Wales is a major aim of the U.K.'s Climate Change Levy.

NATIONAL POLICIES

Many national governments have adopted their own departments or agencies to support the sustainable management of environments and wildlife. In the United States, for example, the Environmental Protection Agency (EPA) was formed in 1970 to repair damage already done to environments and wildlife, and to guide Americans in making a cleaner environment for future generations. The EPA helps the government to enforce new laws protecting environments and wildlife and runs important educational programs to help businesses and individuals adopt better practices.

ENCOURAGING CHANGE

Other governments have targeted specific causes of environmental degradation. In the United Kingdom, for example, the government introduced the Climate Change Levy in April 2001 in an attempt to reduce energy use and greenhouse gas emissions from businesses and industry. A tax (levy) is charged on the energy used in the hope that it will persuade businesses and industries to reduce their energy use (and therefore emissions) as much as possible. The tax can be avoided completely by converting to energy from sustainable sources such as wind and solar energy.

Similar taxes have been used to address other environmental issues in other countries. In Sweden, for example, taxes on the use of pesticides led to a dramatic reduction in their use by farmers, while in France a tax on packaging materials is designed to encourage greater recycling of packaging.

FINANCIAL INCENTIVES

Besides taxes, governments can also use subsidies (financial incentives) to promote more sustainable practices. In Japan and Germany, there are subsidies available for installing solar roofs on new buildings. In the United States a similar program has been used to encourage the greater use of wind power. In England, a Countryside Stewardship Scheme (program) makes payments to farmers who agree to set aside areas of land for the benefit of wildlife.

Grants are also available to farmers who replant hedgerows, or rebuild stone walls — both important habitats for wildlife such as birds and small mammals. By using taxes and subsidies, governments are able to guide their nations toward more sustainable practices, but they do not always work that well. Sometimes the financial incentives are too small to encourage people to change. Another common problem is that people simply do not know about the different programs that might be available.

weblinks

For more information about the Environmental Protection Agency, go to www.epa.org

In the United Kingdom, farmers can now claim payments for restoring and protecting hedgerows. This farmer layers a hedge to encourage thicker growth.

These schoolchildren are being taught about the environment and wildlife they live with in the Korup rainforest of Cameroon, West Africa.

EDUCATION FOR THE FUTURE

One of the most important contributions that governments can make toward the more sustainable management of environments and wildlife is to educate people about the benefits. Many people may be unaware of the impact their actions have on environments and wildlife, and, more importantly, they may be unaware of the alternatives. When people are told about alternatives they may change their practices accordingly. The growth in the use of compact fluorescent lights (CFLs) in homes is a good example of this. Once people realize that they can save energy and money and yet still have the same amount of light as a traditional light bulb, they are normally willing to change. The combined action of individuals changing their behavior can be significant. In North America, for example, the pollution saved by the estimated 275 million CFLs in use

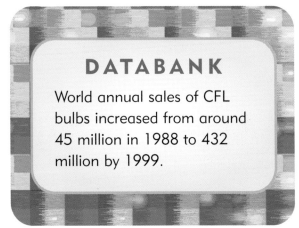

DATABANK

World annual sales of CFL bulbs increased from around 45 million in 1988 to 432 million by 1999.

in 2000, was equivalent to around 3.5 million tons of carbon dioxide. China is one of many countries that is actively educating people about the benefits of CFLs. Between 1996 and 1999, sales of CFLs in China increased by almost 350 percent!

TOMORROW'S ACTIVISTS

One of the most effective ways to safeguard the future of environments and wildlife is to educate the young people of today who will be caring for them in the future. By working with schools and colleges it is possible to train future generations that think and behave more sustainably. Several governments have begun to take this need seriously. In the

The famous flamingos of Lake Nakuru, Kenya, are today breeding again due to an education program that has helped reduce pollution entering the lake from the nearby town.

United Kingdom, for instance, sustainable development is now taught in all schools as part of the new national curriculum. In Kenya, the Kenya Wildlife Service (KWS) is working with young children around Lake Nakuru National Park. Children join a young rangers program and each Saturday they visit the park to learn about and help in caring for the environment and the wildlife living there. In the past, Lake Nakuru's famous flamingo population suffered because of pollution in the water. The KWS hope that by educating local children they will help reduce the amount of pollution entering the lake from nearby Nakuru town.

Sustainable environments and you

IT MAY APPEAR THAT ENVIRONMENTAL PROBLEMS are too large for you and me to deal with as individuals. In reality though, the decisions we make, and the actions we take, can do a great deal to influence the sustainability of environments and wildlife. Some decisions and actions are very direct such as choosing to use environmentally friendly products. Others are more indirect such as choosing to support a particular campaign or group. We can also influence others to consider their own decisions and actions by sharing information with them.

The Centre for Alternative Technology in Wales promotes a sustainable approach to living. The roof solar panels generate their own electricity.

A SUPPORTING ROLE

You or your family may already be involved in supporting environment and wildlife organizations. Most organizations, whether local or global, depend on the support of people like you and me to carry out their work. There are, today, thousands of organizations working to protect and sustain environments and wildlife. Some focus on specific environments such as forests, wetlands, or coral reefs. Others specialize in protecting groups of species (birds, plants, fish, etc.) or

Tourists visit young elephants at an orphan park in Kenya. The orphans' care is paid for by people who sponsor or adopt them to help conserve the species.

particular species such as elephants, mountain gorillas, or pandas. There are also a number of organizations that promote the use of more sustainable technologies such as sustainable energy or organic farming. In fact, whatever your particular interest is there is almost certainly a group that you could contact and become involved with.

The ways in which you can support environment and wildlife organizations extend far beyond purely financial assistance, too. Some organizations offer the chance to adopt animals or parts of endangered environments, while others allow supporters to help as volunteers. In the United Kingdom, for example, the Royal Society for the Protection of Birds (RSPB) used almost 250,000 volunteers, including schoolchildren, to help them conduct a survey of Britain's garden bird population in 2002.

THINK OF THE FUTURE

There are many ways in which you can adapt your daily life to think more about the future of environments and wildlife. Saving energy, conserving resources, recycling, and properly disposing of waste are just a few ways in which you could make a real difference. The most important thing is to think about how the choices you make today might affect the state of the environment in the future.

---

weblinks

For more information about adopting an animal in the wild, go to www.eaaa.net/conservation.htm

---

LOCAL ACTION
Doing your part

You can do your part for sustainable environments and wildlife when you are at home or school. Here are a few examples:

- See if you can reduce energy use around the home.

- Buy environmentally friendly products such as recycled paper or local organic foods.

- Support environmental labeling programs when you go shopping by buying their products.

- Join a local conservation or wildlife club.

- Arrange or participate in a neighborhood clean-up.

- If you have a garden, make it bird-friendly by providing feeding stations or nesting boxes.

- Suggest a school project to survey the state of the local environment and report your findings and suggestions for change.

- Find out more about the issues in this book using libraries or websites.

A vacation based on ecotourism can help conserve future environments. This resort in Jordan provides an excellent example.

TAKE ACTION

There are many things you can do to contribute toward more sustainable management of environments and wildlife. For example, next time your family plans a vacation you could make sure that it is an environmentally friendly one, perhaps run by an ecotourism company. You could even choose a trip that directly helps to conserve environments and wildlife. There are now many such conservation travel programs, ranging from local tree planting projects to tracking endangered wildlife, or surveying threatened habitats. The Earthwatch Institute is one organization that offers such experiences. Since it was founded in 1971, some fifty thousand volunteers have helped its scientists to complete their conservation research and projects.

CONSUMER POWER

You do not need to travel to make a difference — there are many things you can do from your own home. One of our biggest influences is as a consumer. If we choose to buy goods and services that respect environments and wildlife then we can send strong signals to others that they too should take greater care. Many labeling programs now exist to help you make such choices. For example, the Forestry Stewardship Council monitors the cutting and sale of wood to make sure it is done in a sustainable manner. It allows manufacturers who follow its code of conduct to put their logo on their products so that consumers can purchase it knowing it is sustainably produced. A similar program is managed by the Marine Stewardship Council (MSC) to label fish products that come from sustainably managed fisheries.

weblinks

For more information about environmental and wildlife conservation projects, go to www.nature.org

OPINION

'Working with us as a volunteer, not only will you have the experience of a lifetime, you'll also help fund our invaluable work to save the world's wildlife.'

Earthwatch Institute, United States

Community volunteers work with the NatureSearch organization in Queensland, Australia, to survey wildlife in a local wetland.

Sustainable environments and the future

THE TWENTIETH CENTURY WAS the century in which people became aware of the impact their actions were having on the environment. The twenty-first century must become the century in which people cooperate to do something about it. But the twenty-first century will also see human populations increase dramatically from 6 billion at the start to over 9 billion by the end. This increase in human numbers will place further pressure on the Earth's environments and wildlife as the need for land, water, food, fuel, and other resources grows. The pressure will be greater still if the millions who currently live in extreme poverty adopt lifestyles similar to those of people living in more developed countries.

Due to population growth, urgent action is needed to protect the environment for future generations.

TAKING THE STRAIN

There is doubt as to whether the Earth's natural systems will be able to cope with the increased poplation strain expected during the twenty-first century. However, if the way we live our lives changes to better respect environments and wildlife, and live within the limits of the Earth's natural systems, then there is hope for the future. Communities have shown that it is possible to protect natural environments and improve their quality of life at the

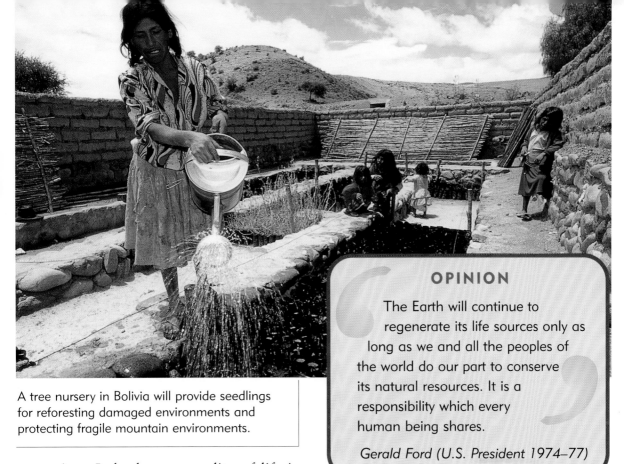

A tree nursery in Bolivia will provide seedlings for reforesting damaged environments and protecting fragile mountain environments.

same time. Indeed, poor quality of life is often closely linked to a poor environment. This means it is in people's best interests to protect and improve their natural surroundings. Many will need help however. For instance, financial help to fund environmental projects such as funding for tree seedlings or for improved waste management facilities. Other forms of help will include education programs to make people more aware of the issues and what they can do to improve matters. Governments will have to work closely with local, national and international organizations, as well as with one another, if changes are going to be effective. Organizations such as UNEP will have to strengthen their role in promoting such cooperation.

IT'S YOUR CHOICE

Ultimately, the future sustainability of environments and wildlife will depend on the choices made by people like you and me. We have the ability to make changes to our own lives and influence others to consider the importance of the issues at stake. If we take that responsibility seriously then a future in which environments and wildlife are managed sustainably is possible.

--- **weblinks** ---

For more information about how you can help to protect the world's environments, go to http://environmentdirectorywildlifeconservation.com

Glossary

Acid rain Produced when pollutants such as sulphur dioxide and nitrogen oxides (emitted when fuels are burned) mix with water vapor in the air to form an acidic solution which falls to the ground. It is damaging to plants, trees, lakes, and buildings.

Agroforestry The growing of trees and agricultural crops alongside each other. The trees shade the crops and can help to improve soil quality.

Biodiversity The variation (diversity) of biological life within an ecosystem, normally measured as the number of different species.

Biosphere The entire Earth's surface, oceans, and atmosphere that is inhabited by living things.

Bleaching A process in which coral dies and turns white (looks bleached) as a result of a warming in ocean temperatures.

Carrying capacity The total population that an ecosystem can support without being damaged.

Conservation To preserve and protect the environment (or any other natural or cultural resource) from change or damage.

Deforestation The removal of trees, shrubs, and forest vegetation. This can be natural (due to forest fires, typhoons, etc) or a result of human action (logging, ranching, construction, land clearance, etc).

Ecological footprint The area of land required to provide a population (such as a city) with the resources they consume and to absorb the wastes that they produce.

Ecosystem The contents of an environment, including all the plants and animals that live there. This could be a garden pond, a forest, or the whole of planet Earth.

Ecotourism Tourism that is sensitive to its impact on environments and local populations and seeks to benefit (or not harm) them by being there.

Emissions Polluting waste products (gas and solids) released into the atmosphere.

Erosion A process whereby something becomes worn (eroded). For example, the removal of material (soil or rock) by the forces of nature (wind or rain) or by people (deforestation, vehicle tracks, etc.).

Eutrophication A process whereby water becomes enriched with nutrients which encourage rapid algae and plant growth. It can result in mats of plant growth which block out the sunlight and starve the water of oxygen, leading eventually to the death of aquatic animals.

Evaporation The process whereby water is converted to a gas/vapor.

Famine A lengthy shortage of food. In extreme cases it can result in widespread starvation and death. Famine can be the result of natural causes, e.g. drought leading to crop failure, or as a result of human interference with food supplies, e.g. during times of war.

Fossil fuels Fuels from the fossilized remains of plants and animals formed over millions of years. They include coal, oil, and natural gas.

Global warming The gradual warming of the Earth's atmosphere as a result of greenhouse gases, such as carbon dioxide and methane, trapping heat.

For further exploration

Ozone layer A layer of gas in the atmosphere that protects the earth from the sun's harmful ultraviolet rays.

Pesticides Chemicals used to kill insects and pests.

Poaching To hunt or fish illegally, usually while trespassing (to be on someone's land without their permission).

PV cells Photovoltaic cells that convert the sun's energy into an electrical current.

Radioactive Substances, such as uranium or plutonium, that emit energy in the form of streams of radioactive particles. These particles are extremely harmful to humans and animals if they come into direct contact with them.

Savannah A dry-land ecosystem dominated by tropical grassland with scattered trees/bushes.

Solar power Electricity generated by converting the energy from the sun, normally using solar panels of PV cells.

Subsidy A payment, normally made by governments, to encourage certain practices.

Sustainable development Developments that meet the needs of today without compromising the ability of future generations to meet their own needs.

Wetlands Areas of marsh or swamp where the soil is saturated with water like a sponge.

Wind power Electricity generated from the wind. Turbines capture wind energy with moving rotor blades and convert it to electricity.

Books

Matthew Chapman and Rob Bowden, *21st Century Debates: Air Pollution*. London: Hodder Wayland, 2001.

Brian J. Knapp, *The World's Environments and Conservation*. Oxfordshire, England: Atlantic Europe Publishing Co Ltd, 1995.

Ewan Mcleish, *21st Century Debates: Rainforests* London: Hodder Wayland, 2001.

Sally Morgan, *Changing Climate*. London: Franklin Watts, 2000.

Malcolm Penny, *21st Century Debates: Endangered Species*. London: Hodder Wayland, 2001.

Index